CLIMATE CHANGE AND GREENHOUSE GASES

CARBON DIOXIDE, THE MOTHERLY GAS

Author

KULDIP CHAND TANGRI

Copyright © 2020 by KULDIP C. TANGRI

All rights reserved. No part of this book may be reproduced or transmitted in any form or by any means without written permission from the author.

ACKNOWLEDGMENTS

I owe to the scientists of the world. Those who established the laws of aerodynamics, motion, and thermodynamics. Without the contributions of these scientists, physics would be a worthless subject where climate science is a 30 % chemistry and 70% physics. I also owe to the Internet, libraries, YouTube, and the people who contributed the information to these sources. And gratitude to Chuck Todd, who unknowingly motivated me to write this book.

Dedication

To the families of the firefighters, those who lost their loved ones fighting wildfires. And the wild animals that got burned alive in the wildfires.

Preface

I have studied auto exhaust and industrial pollution created by the burning or combustion of fossil fuels for the last 49 years. These gases start to deteriorate as soon as they enter the atmosphere, where they mix with the atmosphere's air and lose their strength. Further, they strengthen the water vapor in the atmosphere because these gases are water soluble. Water vapor is not a gas; it is simply one of the forms of water. On the contrary, carbon dioxide, nitrous oxide, nitrogen dioxide, sulfur dioxide, and other related gases are water soluble. These gases are detergent of the atmosphere and nature uses these gases to make sturdy clouds. These gases are an integral part of the climate system.

I have known the environmental and metallurgical scientists of the world were wrong about CO_2, GHG, heat-trapping, and global warming for a long time. Still, I was hesitant to confront them. Since Representative, Alexandria Ocasio-Cortez and Bernie Sanders are critical of fossil fuels, it created an urgency to let the world know the cause and effects of eradication of carbon dioxide (CO_2) and greenhouse gases (GHG) from the atmosphere. Once this book is out, they have an option to accept the truth, either willingly or unwillingly. And world climate science will never be the same.

Thanks to Representative Ocasio-Cortez and Senator Bernie Sanders, I was given a motivation, but it was not enough. On September 15, 2019, I happened to watch *Meet the Press* on NBC, where the show moderator, Mr. Chuck Todd, talked about tropical depressions. Later I ran into another broadcast of Mr. Todd on YouTube on climate change. That was another motivator. About a month later, Miss Greta Thunberg came into the picture. I watched her on TV and felt her pain. The rest is history.

About the Author

Kuldip C. Tangri

Inventor of a hydrogen fuel system for a vehicle

Patent Number 4085709

United States of America

Issue Date 04/25/1978

Filing Date 1975-12-04

Application filed by Kuldip C. Tangri

The author has studied auto exhaust and industrial pollution created by the combustion of fossil fuels extensively for the last 49 years.

Table of Contents

1	This Is Science	1
	1.1 Removal of These Gases	2
	1.2 These Gases are Essential	2
	1.3 Victoria Falls	4
	1.4 Svante Arrhenius	5
	1.5 Heat Trapping	6
	1.6 Ricocheted Sunrays	8
	1.7 CO_2 Is a Straight Line Crooked?	8
2	Nature Depends on Humans	9
	2.1 These Gases Are Part of Our Life	9
	2.2 CO_2, GHG and Atmosphere	10
	2.3 Water Vapor Is Not a Gas	10
	2.4 The Destinations of Water Vapor	11
	2.5 This Chimney	11
3	Anyone Who Accuses Carbon Dioxide	13
	3.1 The Conclusion?	13
	3.2 Improve the Performance	14
	3.3 Engineers and Scientists	14
4	Chuck Todd's Analysis	16
	4.1 NASA Research Associate	16
	4.2 Retrace Our Steps	18
5	Mathematics and Science	20

	5.1	Chemistry, Mathematics, and Physics	20
	5.2	Zero Carbon Footprint	21
	5.3	Heat Transfer	21
6		Materials	23
	6.1	Properties of CO_2 and GHG	23
	6.2	CO_2 and GHG Absorb Heat	24
	6.3	CO_2 Is a Refrigerant of the Atmosphere	24
	6.4	CO_2 and GHG Are Mobile	24
	6.5	CO_2 and GHG Are Water Soluble	25
	6.6	Fortifying the Atmospheric Air	25
7		Clouds and Science	27
	7.1	It Is a Simple Science	28
	7.2	Antifreeze in the Atmosphere	28
	7.3	Acid Rain	29
8		Transportation of Nature	30
	8.1	Without CO_2 and GHG	30
	8.2	Desertification Reduction	31
	8.3	Humidity	32
	8.4	The Sea Levels Down	32
	8.5	Wind Speed and Density	33
	8.6	Glaciers Melting	34
	8.7	Wildfire Reduction	35
	8.8	Earth Is Getting Hotter	36
	8.9	Sources of the Earth's Energy	37
	8.10	TED Talk	38
	8.11	Somebody is Cooking Books	39

9	Volcanoes and Wildfires	40
9.1	The Climate Is Constantly Changing	41
9.2	CO_2 and GHG Are Crucial	42
9.3	Thanks to China and India	42
9.4	Kyoto Protocol	43
10	Greenhouse Gases (GHG)	44
10.1	Carbon Dioxide (CO_2)	44
10.2	Methane (CH_4)	46
10.3	Nitrous Oxide (N_2O)	46
10.4	Nitrogen Dioxide (NO_2)	47
10.5	Sulfur Dioxide (SO_2)	48
11	Atmospheric Layers	49
11.1	The Troposphere	50
11.2	The Stratosphere	51
11.3	The Mesosphere	51
11.4	The Thermosphere	52
11.5	The Exosphere	52
12	Clouds	53
12.1	Types of Clouds	53
12.2	Low-Level Clouds	53
12.3	Stratus Clouds	54
12.4	Stratocumulus Clouds	54
12.5	Arcus Clouds	54
12.6	Mid-Level Clouds	54
12.7	High-Level Clouds	54
12.8	Cumulonimbus Clouds	55

	12.9	Pyrocumulus Cloud 55
13		The Intergovernmental Panel 56
	13.1	The Pseudoscience 56
	13.2	Heatsinks 57
	13.3	Suitability of Material 58
	13.4	Parchment Paper 59
14		Adam the Project Engineer 60
	14.1	Properties of Air 61
	14.2	Water Vapor 61
	14.3	The Climate 62
15		Janicki Omni Processor 63
	15.1	Nature Is a Mighty Distiller 63
16		Abbreviations 64

This Is Science

Climate change and global warming are the result of the deficiency of carbon dioxide (CO_2) and related greenhouse gases (GHG). The World Meteorological Organization (WMO), established by the United Nations Organization (UNO), the Intergovernmental Panel on Climate Change (IPCC), the Kyoto Protocol, the United Nations Framework Convention on Climate Change (UNFCCC), the Environmental Protection Agency of America (EPA), the National Aeronautics and Space Administration (NASA) and their fellow scientists are responsible for the unintentional worldwide disaster of the climate system. Just in the name of heat trapping. They have eliminated multi-trillions of gigatons of these gases from the world atmosphere in the last 50+ years and continue to do so with an iron fist. This drastic removal of CO_2 and GHG from the Earth's atmosphere has weakened nature's transportation system. Nature distributes nutrition and other ingredients to where they belong 24/7 by means of these gases. These gases help the atmosphere air to transport snow to the highest peaks of the mountains and water supply to the springs, the water holes of African jungles, and around the world. There are thousands of things nature does, where we do not know how and why. This is one of them. The deficiency of CO_2 and related GHG (gases produced by the combustion of fossil fuels that are water soluble) has created a road to the devastation of the ecology, economy, and weather system of the world. Water is accumulating in the seas, and sea levels are rising as seashores are shrinking. Desertification and deforestation are increasing due to the wildfires and droughts. CO_2 and GHG do not produce or multiply energy (heat) like hydrogen-3 (3H) does. Therefore, these gases have nothing to do with global warming.

Removal of These Gases

Removal has made the atmosphere air thiner and lighter, causing the speed of wind upsurge. This effect alone has caused the loss of human lives and multi-trillions of dollars lost all over the world. One of the recent tragedies was Hurricane Dorian hit the Bahamas in 2019. It was a category 5 and the wind speed went over 183 miles an hour.

These Gases are Essential

They are essential components of the climate system in the atmosphere. CO_2 and GHG are a refrigerant of the Earth, and scientists know that CO_2 and related GHG absorb heat; that is what the refrigerants do. They absorb heat from the surface of the subject (like the evaporator in a refrigerator) and cool it mechanically. In this case, the subject is Earth, and CO_2-rich atmospheric air is a refrigerant, which absorbs the heat from the surface of the Earth and cools this air by convection.

Climate Change and Greenhouse Gases

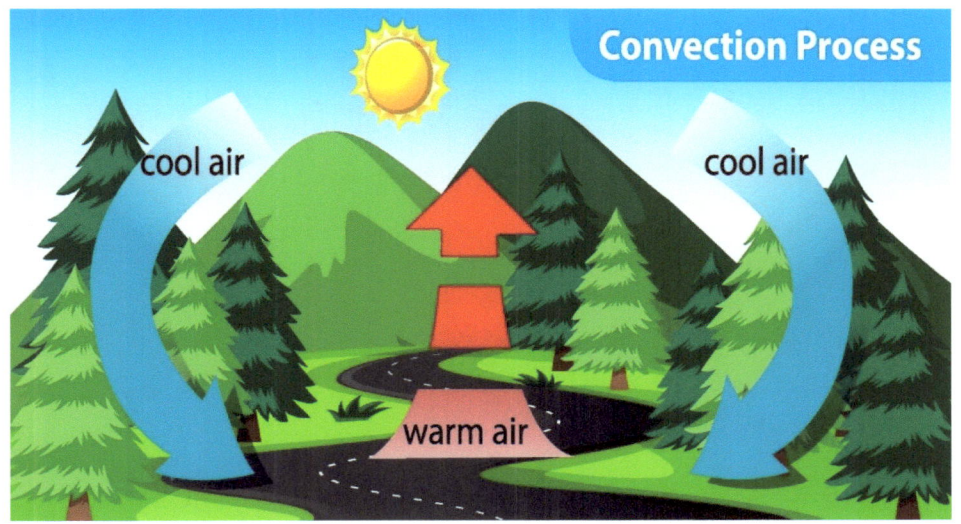

Figure 1

Therefore, the Earth's atmosphere needs these gases to cool the Earth. By increasing the percentage of these gases in the atmosphere air is going to increase the conduction capacity. Which is going to boost the capturing capacity of the heat from the surface of the Earth. And cool the Earth faster by convection. The convection cooling system works as in the above Figure 1, and it works exactly as shown. It is going to be explained further in this book.

Multi-trillions of trees produce CO_2, shed leaves, and produce firewood. Do you really think that the universe, which created these trees and made them shed their leaves yearly and produce CO_2 all the time, is stupid? Fifty-five years ago, people were living on the foothill of the Himalayas. During the fall, they accumulated leaves. Throughout the winter, they burnt those leaves to use them for cooking and warm up their residents. I believe some of them are still doing it. In India and neighboring countries, people used to burn coal firewood and animal dung for cooking and other heating needs. Nowadays, most people use natural gas for daily cooking and other domestic needs. It means nature wants us to produce these gases.

Victoria Falls

Victoria Falls is a recent tragedy of the deficiency of CO_2 and related GHG. On February 19, 2020, I was watching good morning America on ABC (American Broadcasting Company), where I found out about the Victoria Falls and I researched it. This fall is the widest curtain of falling water in the world, and one of the Seven Wonders of the World. This fall and wildlife attract hundreds of thousands of tourists every year. This waterfall divides Zambia and Zimbabwe. Lately, the water sources of these magnificent falls are dwindling, and it dries up often, and the situation is getting worse. The droughts and delayed rainy season often affect the waterfall, but it is still one of the most magnificent waterfalls in the world. If the world did not confront the pseudo-scientists and WMO did not correct their scientific mistake, the world could lose this waterfall in entirety. The water sources of this fall are a lifeline of Zambians. There were times when the water levels of rivers dropped, and it affected hydropower and caused a power shortage in Zambia. People, businesses, and wildlife suffered a lot. Droughts are not only affecting Zambia; they also affect thousands of other countries, states, and municipalities worldwide. Those dependent on hydropower. The President of Zambia blames the doubters.

Svante Arrhenius

Svante Arrhenius, a Swedish scientist, presented a paper about the effects of carbonic acid in the air in 1895. He highlighted the impact of CO_2 on the temperature on Earth. At that time, the Industrial Revolution was on the rise. Mr. Svante Arrhenius pointed out that CO_2 produced using fossil fuels may be beneficial. He further added that CO_2 might stabilize the climate of the Earth, stimulate plant growth, and increase food production. He did not explain how and why. Somewhere on the way, Mr. Svante Arrhenius's message about the CO_2 got twisted. This may not be unintentional, but the consequences are lethal. I believe the United Nations (UN) was the lead player, who emphasized that greenhouse gases cause global warming and heat trapping. And the world obliged. Even though it is not true, as this myth went unchecked, it got amplified. The UN added the World Meteorological Organization (WMO) in 1950. There are a few more intergovernmental agencies such as the Intergovernmental Panel on Climate Change (IPCC). And there followed a further extension of the Kyoto Protocol. It didn't stop yet and it is still going on. This action has taken place in the last 70 years. In 1970, late President Richard Milhous

Kuldip C. Tangri

Nixon created the Environmental Protection Agency (EPA), and he told the world that he would be the environmental president. Because he wanted a clean environment. Richard Milhous Nixon lost his political battle, but EPA thrived. It is the EPA that ensured the United States of America has a clean environment. Somewhere on the way, the EPA became a follower of the UN instead of an individual organization. I do not know how the WMO scientists started to blame the CO_2 and related GHG for global warming. Could it be because CO_2 absorbs, disperses, and radiates heat?

Heat Trapping

The fact that CO_2 and GHG are trapping heat is an absurd allegation. From where is the heat trapping coming? Large groups of people are making a fool of 7 billion people. The future scientists should ask their teachers: is the Earth's atmosphere a glasshouse? How do the greenhouse gases trap the heat? What happens to the trapped heat when the wind blows even five miles an hour? Please explain how CO_2 causes global warming. What does the heat absorption of CO_2 and GHG have to do with global warming? CO_2 and GHG do not have anchors to hold them down in one place, and they do not have directional valves, which would allow the flow of cold or heat in one direction. These gases conduct cold or heat constantly in all directions. During this process, they gain or lose heat energy. CO_2 and GHG are not trappers of heat or cold; they are conductors, disperses, and radiators of heat or cold. These gases are not immune to the laws of thermodynamics, where they cool the Earth by conduction, convection, and radiation. They move in all directions to equalize the atmospheric temperature. During this process, they gain or lose heat.

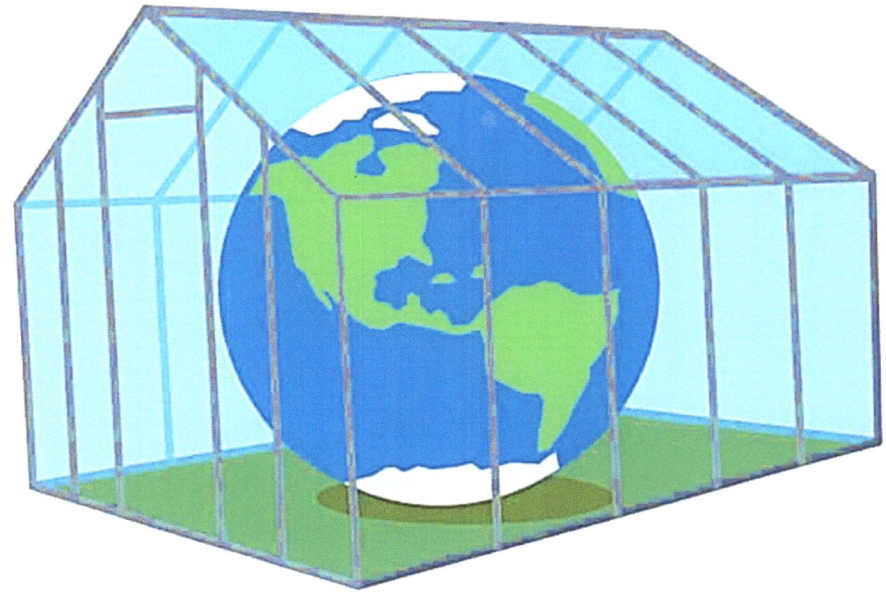

Figure 2. Copied from the website of the National Aeronautics and Space Administration (NASA).

In Figure 2, NASA's writer is explaining how the greenhouse effect works. Of course, the greenhouse is a building made of glass with glass walls and a glass roof and has a heating source to keep it warm. They are used to grow plants in them. The inside of the greenhouse is isolated from the outside atmospheric temperature. The inside of the greenhouse is going to stay warm as long as all the above conditions are in position. The only problem is that the atmosphere of Earth is not enclosed glasshouse; there is wind blowing all the time, and the wind speed is controlled by the air density and the atmospheric air pressure differences. Sometimes the wind speed goes over 100 miles an hour. The atmosphere of Earth does not have walls and roof, not even a tarpaulin to cover it. Therefore, there is no comparison. Scientifically CO_2 and GHG do not trap heat – they conduct. Remember, heat always travels from hot to cold.

Ricocheted Sunrays

All objects, including human bodies, absorb and emit radiation, and the strength of radiation depends on the temperature of the object. The same goes for heat radiating from the Earth. Sooner sun rays hit the Earth, and the Earth absorbs sun rays instantly. Any of the ricocheted sun rays are too weak to carry any heat and go over 20 feet (6.1 meters) high into the atmosphere. It is like a golf ball. If you throw a golf ball on a hard floor, it is going to bounce, but it will not on grass. However, reflective material can deflect sun rays and provides some protection, and the sun rays can ricochet a few feet farther. It is like heat radiating from a bonfire, where the heat reaches only a few feet to a few yards around the bonfire.

CO_2 Is a Straight Line Crooked?

One of the quotes of Alfred Nobel: "Lawyers have to make a living and can only do so by inducing people to believe that a straight line is crooked." I believe climate change is for real, but blaming it on CO_2 is a straight line being addressed as crooked line.

Nature Depends on Humans

Nature depends on humans to burn fossil fuels to produce CO_2 and related GHG. The Industrial Revolution was beneficial to the world. Lately, the CO_2 and related GHG supply have been dwindling. Therefore, nature is not getting enough CO_2 and GHG to meet her daily needs for proper function. Nature has a harmonious way of getting CO_2, as the trees produce CO_2 at night, and we the people use fossil fuels.

Civilization and wildfires have eliminated multiple trillions of trees, which has affected the supply of CO_2. Seas produce nitrous oxide (N_2O), but the Earth is so hot that seawater is evaporating faster than the sea can produce N_2O to accommodate the water vapor. It looks like nature's harmonious ways are not producing enough CO_2 and GHG. On top of that, all nations are on the zero-emission bandwagon. To meet her daily needs, it doesn't leave nature with too many choices. The first option is wildfires; look around at the number of wildfires around the world. The next option is volcanoes. Active volcanoes produce about 600 million tons of CO_2 yearly. There a few volcano eruptions in the works.

These Gases Are Part of Our Life

We are surrounded by these gases 24/7. I think this is the way nature distributes and regulates these gases to surround us. Otherwise, nature uses volcanoes and wildfires to produce these gases violently and in an extremely large quantities.

CO_2, GHG and Atmosphere

A certain percentage of CO_2 and other GHG fortify the atmospheric air. Where the atmospheric air is a major prime mover, this action gives it (the prime mover) the ability to carry and transport the water vapor. These gases alter the freezing point of water vapor in the atmosphere to prevent it turning into a downpour, further make the healthy clouds. They are all integral part of the climate system. Once these gases mix with water vapor and become clouds, there is no going back; they are consumed. That is why the atmosphere of the Earth needs millions of tons of CO_2 and other GHG daily to function.

Water Vapor Is Not a Gas

It is one of the three forms of water. These forms are ice, water, and vapor. Water continually evaporates, even below subzero temperatures. Ice is converted into vapor without converting into water; this is called sublimation. It occurs all the time, and it reduces the glaciers drastically. Water also evaporates either from a pond, river, or sea. It cools the Earth. Water vapor is a byproduct of cooling of the Earth. The multi-trillion tons of water currently in the form of vapor in the atmosphere needs to be removed and transported away from the site of vaporization. Atmosphere air alone cannot remove it. Atmosphere air can move it around a few miles, but eventually it is going to condensate and turn into a downpour. To avoid the downpour as mentioned above in CO_2, GHG and atmospheric water vapor is fortified with CO_2 and GHG to decrease the freezing temperature of the water vapor and turn into healthy clouds to transport to other destinations.

The Destinations of Water Vapor

These include fields, lakes, mountain peaks, ponds, rivers, streams, and water holes of the world – even the water sources of Victoria Falls. Remember, CO_2 and GHG are added to the atmosphere air to sustain the water vapor in the atmosphere, by altering (lowering) the freezing temperature of the water vapor in the atmosphere to make the healthy clouds.

This Chimney

The atmosphere of Earth acts like a chimney – specifically a troposphere. Heat and smoke from the Earth, wildfires, volcanic dust and gases, water vapor, etc. Pass through this chimney to get to a higher altitude. All these items leave the Earth to cool the Earth. They do not cause the Earth to warm; they are the byproduct after cooling the Earth.

Kuldip C. Tangri

Therefore, please do not blame the CO_2 and all the GHG for global warming. Actually, they are the tools to accommodate the Earth's cooling and they stabilize Earth's temperature. Any deficiency of these gases causes the Earth to warm up. Of course, excess of everything is bad.

Anyone Who Accuses Carbon Dioxide

Anyone who accuses carbon dioxide (CO_2) of global warming is not a scientist! With due respect to the scientists of the WMO and all the affiliated organizations for climate change and global warming, please explain this to the 7 billion people in the world – those who have put their lives on hold because they trusted you. They are young mothers, the mothers to be, and the mothers who wanted to be. They want to know what CO_2 has to do with climate change and global warming. Their lives have changed forever because all of you have promoted information which is not proven scientifically. I believe you knew that this information was not adding up scientifically. Either intentionally or unintentionally, you have misguided the world, and you are responsible for the destruction of the world climate system and global warming. You blamed CO_2 and greenhouse gases for climate change and global warming, which is not true. CO_2 is not black dust or soot. It is with us from the day we are born, 24/7. It runs through our veins. We sleep with it, and we wake up with it. All trees produce it. This is a motherly gas.

The Conclusion?

How did you conclude that CO_2 is the problem? Did you use the laws of aerodynamics? Did you use the laws of thermodynamics? Maybe any of the laws of Isaac Newton? Any physics or chemistry? I know you have used a part of the science to identify the properties of a material. Otherwise, you would not know that CO_2 and GHG absorb heat. But you left a lot more properties of CO_2 and GHG out. CO_2 has a low freezing point and is used to preserve medical and food items.

Improve the Performance

The universe comprises millions of different materials. They all have different properties, and all these materials can be further compounded with each other to improve their performance. That is what CO_2 and GHG do: they combine with atmospheric air and improve the performance of that air to conduct heat.

Engineers and Scientists

You, the climate engineers and scientists, and your predecessors, either did not know this part of the science or overlooked it. Now is the time for you, the scientists, to go out and tell all the children of the world, who are terrified to death, that CO_2 is safe. Also tell them, "we are sorry that we dropped the ball in the matters of climate science." Also tell all the coal miners who have lost their livelihood forever. Further, tell the families of the firefighters who lost their loved ones fighting wildfires (Figure 3) and the wildlife who got burnt alive in the wildfires. Now is the time to come clean. Also have mercy on the young Swedish girl Greta Thunberg.

Please let her know that CO_2 and GHG are safe, and she can breathe a sigh of relief.

Figure 3 Pixabay

Chuck Todd's Analysis

On September 15, 2019, I was watching *Meet the Press* on NBC. Chuck Todd, the moderator of the show, pointed out the increasing numbers of **tropical depressions** (TD). He gave the numbers of TD per decade. Starting in the 1980s, there were 93 TD; in the 1990s, there were 110 TD; in the 2000s, there were 151 TD; and in the 2010s, there were 149 TD, and more are to come. Also, there was a further analysis of the costs and the number of human lives lost due to these disasters. I liked the analysis and feistiness of Mr. Chuck Todd and wanted to see who this person was. So, I went over to Google to look him up. While I was looking up his bio on Google, I saw a big heading: "Chuck Todd Refuses to Give Airtime to Climate Deniers. The Science Is Settled." I had an urge to read that article, so I clicked on it. It was better than I expected.

NASA Research Associate

On December 31, 2018, HuffPost posted an article by journalist Amy Russo. In the article, there was a video discussion about climate change. Mr. Chuck Todd, the moderator of the show, started with a question to Dr. Kate Marvel, an associate research scientist with the National Aeronautics and Space Administration (NASA), about climate change. Here is Dr. Marvel's point of view on climate change. Exactly in her own words: "Oh my gosh… I wish I knew… I wish I had a good answer for this because as scientists, what we want to do, what we're always tempted to do, is show more data and more graphs like there's gonna be some magic equation that's gonna convince everybody... and there isn't, you know... I don't think that a lot of the reluctance to accept climate change... *I don't really think that's about the science...* I think that's about values... I think that's about the sort of deep story of how people see themselves, so I think it's really important for scientists to go out in communities... engage with what's important to people in communities." Next, *NBC News* chief environment correspondent Anne Thompson gave her opinion on the effects of fossil fuel burning on climate change. These are her exact words: "... and that's important because I always liken the climate change to cancer... they're both such huge issues... they're really hard to get your head wrapped around it, if you will, but if you look at pictures... take a trip to Glacier National Park out in Montana in 1850 when the Industrial Revolution started, we started burning coal and sending greenhouse gases in the air... there were 150 glaciers in that National Park; today there are 26, and they're in danger of losing those 26; they're really threatened."

    My observation of the entire broadcast was that Dr. Kate Marvel said, "We do not have any scientific data to support our claim." However, she still wanted the experts to go out into communities and talk to people to convince them about the negative impacts of fossil fuels. **This meant going out and lying to people**. Then Anne Thompson spoke, further muddying the waters when she started talking about Glacier National Park in Montana. There were 150 glaciers in 1850, at the beginning of the Industrial Revolution, but there were only 26 glaciers as of December 31, 2018 (date of the broadcast). It seemed obvious that Ms. Thompson wanted to exaggerate fossil fuel usage and the reason for the melting glaciers. Why didn't she start in 1970 instead of 1850? Because glaciers were doing well in 1970. If Ms. Thompson began in 1970, then she could not make her point that CO_2 and GHG are the culprits of global warming. There were three other panelists on the show: an elected official who was concerned about rising sea levels and two other government officials. Those three were telling, their personal on-the-job experiences. Dr. Marvel said she did not have scientific facts, but wanted to continue to emphasize it, and Chuck Todd said yes, it is overwhelming, and he suggested a carbon tax. Link to the video: https://youtu.be/TDSsU5kukZc

Retrace Our Steps

Previously in chapter 4, Chuck Todd pointed out that with each decade, the numbers of TD went up. To find the reason for these incremental increases in TDs, we should retrace our steps. We are going to find out that all the subsidiaries of the United Nations and the Environmental Protection Agency (EPA) of the USA have eliminated multi-trillions of megatons of CO_2 and GHG from the atmosphere of the Earth since 1980 and continue to do so with an iron fist. Do you think that the drastic removal of CO_2 and GHG from the atmosphere by the above agencies would not have any effect on the climate? The incremental increase is the result of the drastic removal of CO_2 and GHG from the atmosphere of Earth. It also led to an increase in the temperature of the Earth and weakened the performance of the global weather system. Therefore, Mr. Chuck Todd, ***The Science Is Not Settled!***

Mathematics and Science

If it does not add up, it is not science. It is time for you, the scientists, and the students of the world's academies and universities on the climate and environmental science, to reexamine your position on CO_2 and related GHG. A real scientist should be able to understand and control the world temperature to a certain extent by managing these gases. The time has come when a scientist should be able to tell the world why the peak of Kilimanjaro is losing its Snowcap.

The world scientists should have all the answers. How can we replenish the mountain peaks with snow? How can we reduce downpours, and how can we bring the sea levels down and reduce the number of wildfires? How can we decrease desertification and mitigate the droughts? Oh! Yes, and how can we reduce the speed of the wind? Some answers to these questions will be answered in this book.

Chemistry, Mathematics, and Physics

All climatology, environmental, and meteorological science consist of chemistry, mathematics, and physics. There is zero room for mistakes in getting the right answer in these subjects. There are thousands of scientists that graduate every year from the universities of the world after studying the above subjects. I believe these universities have a way to manipulate the science or the calculations when it comes to CO_2 and other GHG produced by fossil fuels. Otherwise, scientists would have known the benefits of CO_2 and other GHG in the atmosphere. How could these students graduate and be satisfied with the results in the matter of CO_2 and other GHG?

Zero Carbon Footprint

I know you, the scientists or faculty, use the above scientific knowledge to calculate the carbon footprint. But they do not offset the consumption of CO_2 and related GHG in the atmosphere. Because these gases are water soluble, once they enter the atmosphere air, they lose their concentration. They further mix with the water vapor, already in the atmosphere, and lose their identity. CO_2 is no longer CO_2, and it becomes a trace of carbonic acid in the clouds. Once these clouds turn into rain or snow, they get washed down. Therefore, there is a zero or negative footprint in the atmosphere! These gases are the detergent of the atmosphere. They take the dust particles and germs out of the atmospheric air, wash them down in the form of rain or snow, and leave the atmosphere fresh and clean.

Heat Transfer

This is fourth-grade science!

According to the laws of thermodynamics, heat is transferred by conduction, convection, and radiation, as you can see in the Figure 4 below. The radiation of the burners heats the pot, and the pot heats the liquid by conduction. Then the convection cools the liquid. The handle of this pot gets hot by conduction. Heat always goes from hot to cold, and there are no exceptions.

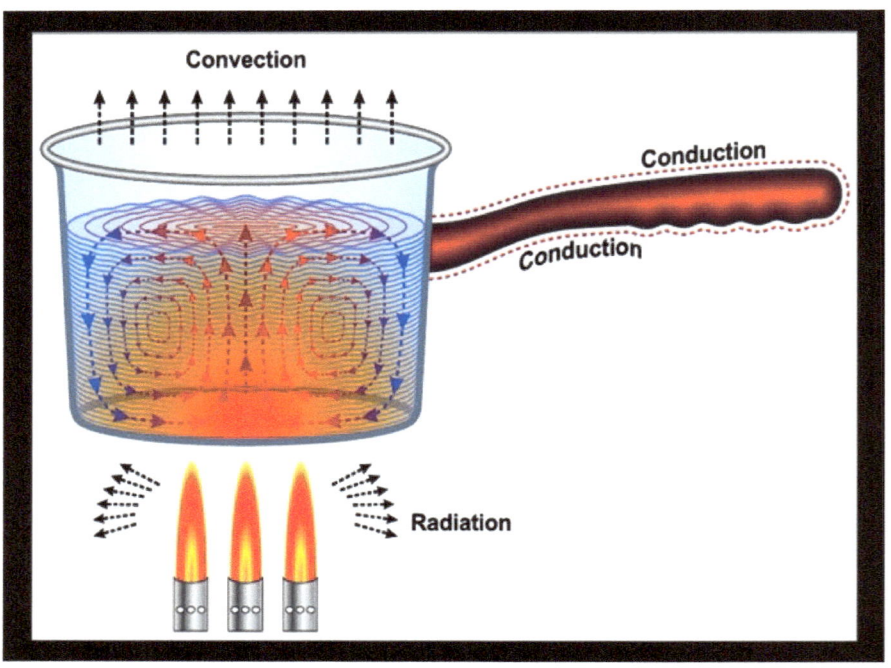

Figure 4

In the above Figure 4, the burners are heating the pot by radiation. The pot heats the handle of the pot and the liquid in the pot by conduction. Convection and evaporation cool the liquid. The liquid in the pot is an integral part of the cooling system of the pot. Without the liquid in the pot, it would get overheated or burnt.

Materials

One cannot elaborate on the function of a material when a person does not know anything about what a material does. I am talking about GHG created by burning or combustion of fossil fuels. Specifically, CO_2. All materials, air, CO_2, GHG, liquids, and solids have thermophysical properties. These properties are thermally conductive. They absorb, disperse, and radiate heat according to their molecular identity. All materials, including air, gases, liquids, and solids, thermally conduct according to their molecular identity. All gases occupy the atmosphere, according to their molecular weight and properties.

Properties of CO_2 and GHG

CO_2 and GHG have three unique properties, which are heat absorption, mobility (by blending with atmospheric air), and water solubility. Without these three properties of gases, the climate system cannot work. CO_2 and GHG have the properties of liquids and solids in the heat absorption field. Because, they are gases this gives them an advantage over the liquids and solids. This is a unique property, meaning they can blend with the atmosphere, air and go wherever it takes them. Compared to liquids and solids, they need physical transportation. The climate system of Earth cannot work without these gases. Atmospheric air is just a destructive force without these gases.

CO_2 and GHG Absorb Heat

CO_2 and GHG's absorb and disperse heat, and that makes them important. They absorb heat from all directions and disperse it in all directions. This is a vital tool for the climate system to transport heat or cold to equalize the atmospheric temperature. Otherwise, the Equator will be hot forever and the North and South Poles will be cold forever.

CO_2 Is a Refrigerant of the Atmosphere

CO_2 and GHG work as refrigerants in the atmosphere. In a refrigerator, a refrigerant absorbs the heat from the surface of the evaporator of a refrigerator, and a mechanical device brings the hot refrigerant out to cool it. In nature, CO_2 and GHG help the atmosphere to capture the heat from the surface of the Earth and cool the Earth by convection, naturally and automatically.

CO_2 and GHG Are Mobile

CO_2 and GHG have the properties of liquids and solids. Compared to liquids and solids, however, CO_2 and GHG have an advantage over both liquids and solids. CO_2 and other GHG fortify the atmospheric air and become a part of that air. They go where the atmosphere, air goes.

CO_2 and GHG Are Water Soluble

Unlike noble gases, they are water soluble; therefore, their life cycle in the atmosphere is limited. This means nature uses them constantly. As soon as they leave the production source, they begin to deteriorate. They lose their identity and strength by mixing with the atmospheric air simultaneously with the water vapor in the atmosphere air. It is one of the vital properties of CO_2 and other GHG. This action increases the atmospheric air to carry the water vapor farther in distance and higher in altitude, by decreasing the freezing point (temperature) of the water vapor to avoid it turning into a downpour, but to make healthy clouds. Those can get to the mountain peaks without falling apart. They also reach the vegetation and organisms in the form of air, rain, and snow. The vegetation and organisms need these elements. All GHG, including CO_2, nitric oxide (NO,) nitrogen dioxide (NO2), nitrous oxide (N2O), and sulfur dioxide (SO2) are a part of nature's delivery system.

Fortifying the Atmospheric Air

Scientists know that pure atmospheric air does not absorb heat. Without this property, it is worthless in the matter of heat transfer and the climate system. It can get cold or hot, but it cannot absorb cold or heat to transport, because it does not capture (absorb) or disburse any heat. Nature uses CO_2 and other GHG to mix with atmospheric air to increase the heat absorption capacity of the atmosphere air. In this way, this fortified air can absorb and disperse heat while carrying heat or cold wherever it goes. The carrying capacity of this fortified air depends on the percentage of CO_2 and related GHG in the atmosphere. Therefore, this fortified air is nature's transportation system of heat and cold.

Clouds and Science

We often hear on the news about downpours, floods, and record water-dumping rain. This is the result of not enough CO_2 and related GHG in the atmosphere to fortify and sustain the water vapor in the atmosphere. Water vapor in the atmosphere can condense and turn into a downpour or rain anytime. To carry water vapor from the oceans to the peaks of mountains, there is a science to it. The atmosphere needs three factors to make clouds and get them to their destination. Those three factors are transportation, water vapor, and something to hold the water vapor during transportation. Nature has atmospheric air, a powerful transportation, and there is enough water vapor to transport. Water vapor does not stay in the atmosphere by the grace of God. There is a science to keep the water vapor in the atmosphere until it reaches the destination. That is where CO_2 and GHG come in. As we know, the boiling point of water is 100°C (or 212°F). Therefore, water vapor in the atmosphere can turn into a downpour anytime.

Nature fortifies atmospheric air with CO_2 and other GHG to maintain the water vapor in the atmosphere. This fortified air not only transports heat or cold, but also maintains and carries a higher amount of water vapor in the form of clouds and humidity at a lower temperature. It also fortifies the clouds so they can reach a higher altitude and travel farther without falling apart in the form of downpours. These gases mix with water vapor in the atmosphere because they are water soluble. This action of greenhouse gases lowers the dew point (freezing) temperature of water vapor in the atmospheric air. These gases keep the water vapor in a semi-gaseous form in the atmosphere, even below -40°C.

It Is a Simple Science

As we know, the freezing temperature of pure water is 0°C (or 32°F). Now let us take the freezing temperature of CO_2, which is -78.5°C (or -109.2°F). When CO_2 mixes with water vapor in the atmosphere, it reduces the freezing (dew point) temperature of the water vapor considerably. Each GHG has a different freezing point, so they can reduce the freezing temperature of the water vapor accordingly, whereas the freezing temperature difference between CO_2 and water (-109.2 + 32) is a 141.2°F temperature difference. Therefore, the higher the amount of CO_2 in the atmosphere, the lower the freezing temperature of water vapor is going to be.

Antifreeze in the Atmosphere

We have many choices of antifreeze to lower the freezing temperature of the liquids on the surface of the Earth, either liquids or solids, such as ethylene glycol, hybrid organic acid, inorganic additive, and organic acid. We used to add sodium chloride to water to reduce its freezing temperature, and we spread salt on roads to melt the ice because salt lowers the freezing temperature of water. The climate system also needs antifreeze to treat water vapor in the atmosphere. The only option is CO_2 and other GHG. The salt does not come with the water vapor from the sea; it stays behind. I know some of you say salt is part of the clouds – not so. It can only enter the clouds in the form of dust, which is nearly impossible. Sometimes, there are dust clouds in the atmosphere, but if there is enough humidity in the atmosphere, it will reduce the frequency of the dust clouds.

Acid Rain

When we talk about CO_2 and GHG, we must discuss acid rain. Acid rain is possible where industrial areas are releasing large quantities of acidic gases. Nowadays, regulations are so strict it is out of the question. It may be possible in the regions where sulfur mines and volcanic eruptions are common, however. All we can do is keep the Earth cool, so volcanoes stay dormant. CO_2 and GHG are naturally acidic, and nature has used them for millions of years to make healthy clouds. They are a required component to reduce downpours and heavy rainfalls. I read an argument about turning the sea acidic – not so. The only way sea can turn acidic is if the Earth gets so hot, it blows up in the sea and releases the gaseous pressure in the sea like a tsunami. If you lived through the sixties, you know our atmosphere can take a lot more CO_2 and other GHG than the current government regulations allow.

Transportation of Nature

The atmospheric air is a powerful prime mover, like a train engine. As mentioned previously, CO_2 and GHG absorb heat, and they are water soluble. Therefore, they mix with the water vapor in the atmosphere and become a part of the air. They act like boxcars in the atmosphere. This blend equalizes higher magnitude of temperature, also contain a higher amount of water vapor in the form of clouds and humidity, at a lower temperature. This fortified atmospheric air runs on autopilot. It travels in all directions and gets loaded and unloaded quickly and automatically. It captures and releases the heat or cold fastest it can. The centrifugal force of Earth and a pressure difference keep the atmosphere air moving relentlessly. It never stops. It strives to equalize the atmospheric pressure and temperature, and it also transports clouds to their destination.

Without CO_2 and GHG

Without CO_2 and GHG, atmospheric air is lighter and can become high-speed, damaging wind. It can uproot trees and tear apart buildings. An example: Hurricane Harvey in 2017 in Puerto Rico, Texas and the surrounding areas; and Hurricane Dorian, in 2019 in the Bahamas. Damage includes human lives, animal lives, and property worth billions of dollars.

Desertification Reduction

All the problems in the atmosphere have a common solution. If we correct the deficiency of CO_2 and other GHG, everything is going to fall into place. The density of clouds will start to go normal and the transportation system of nature will start to work correctly. It will reduce the downpours and make the clouds sturdier. Those clouds will travel farther and higher in the altitude without falling apart. They will reduce the droughts and desertification considerably. If you ever visit a drought-stricken area, look in the sky, and you will see no clouds and no humidity. See the bottom Figure 5 a tragedy of deficiency of the CO_2 and other GHG. There used to be plenty of water in this area.

Figure 5 Pixabay.com

Kuldip C. Tangri

Humidity

The average relative humidity in the atmosphere is less than 5%; it varies with the geographical area and altitude. A 5% relative humidity is low, and the desired relative humidity in the atmosphere is 50%. Atmospheric water vapor quantity decreases with altitude relative to the surface of the Earth. To increase the relative humidity at a lower temperature, increasing CO_2 and GHG in the atmosphere will do the magic, where the atmosphere can reap the benefits of having enough humidity without raising the temperature of the Earth's atmosphere. There is NASA and **Japan Aerospace Exploration Agency (JAXA)** established a **Global Precipitation Measurement** (GPM). I do not think if they know that **global precipitation can be controlled to a certain extent by controlling the** quantity of CO_2 and other **GHG in the atmosphere.** Do you know honey bees need 50–60% humidity? No wonder honeybees are dying or leaving the country.

The Sea Levels Down

NASA may have noticed that clouds are thinning in the Earth's atmosphere since late 1970 During this period, what did the WMO do? Reduced billions of tons of CO_2 and related GHG from the Earth's atmosphere. "In 2014, global sea level was 2.6 inches above the 1993 average." (This info is obtained from the National Ocean Service website). At any given time, there should be trillions of tons of water vapor held in the atmosphere in the form of clouds and relative humidity. That is not currently transpiring, because the scientists ignored the science and continue to reduce CO_2 and related GHG in the atmosphere, where CO_2 and GHG would help the atmosphere to hold a higher amount of humidity at a lower temperature. Once the atmosphere has enough CO_2 and GHG, the relative humidity level in the atmosphere will start to go up. Presently, the average humidity in the atmosphere is about 5%, and the desirable humidity is about 50%. Further, clouds will start to reach their destination. The number of downpours will decrease. Eventually, the sea level will drop drastically.

Wind Speed and Density

Wind speed and density have a direct correlation. Where it is explained in **Sir Isaac Newton's laws perfectly.** Drastic removal of CO_2 and GHG has made the air thinner and lighter. During Hurricane Harvey, in addition to the wind damage, Texas and Louisiana had catastrophic flooding and many deaths. Because Harvey stayed in one place for days, since the air didn't have enough density to move or break Hurricane Harvey apart, it was like a turbine running without an impeller. Similarly, in Australia, winds often uprooted trees. There is no need to do anything to further slow down the wind speed. Once the atmosphere has enough CO_2 and related GHG, the atmosphere air will get thicker, and the wind speed will slow down due to the weight. The relative humidity and clouds will put extra weight in the atmospheric air, and things will fall into place.

Glaciers Melting

Glaciers melting causes sea levels to rise is somewhat true. It may seem obvious that CO_2 and GHG are causing the global warming and that global warming is melting glaciers, which cause sea levels to rise. However, it is not true. Actually, it is the result of the deficiency of CO_2 and GHG in the atmosphere air. These gases in the atmosphere reduce the direct exposure of the sun's rays on glaciers and also reduce the sublimation due to the increase in the density of the atmosphere air. CO_2 may provide an invisible protective coating of chemicals around glaciers, so they can still deflect the sun's rays. Once enough CO_2 and GHG are in the atmosphere, the melting of glaciers will slow down, because the air will get thicker and slower and may work as a protective coating or thin layer of insulation, like a cooler.

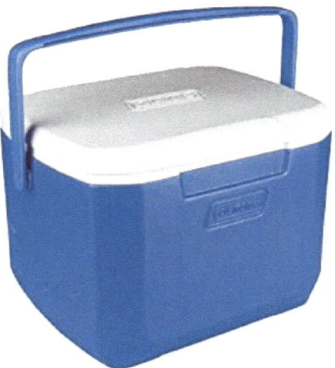

Figure 6 Coleman coolers

We use coolers as in Figure 6 to keep food hot or cold, and these coolers do not produce any energy to help the cause. All they do is isolate the inside of the cooler from the outside temperature. Further, increased humidity in the atmosphere will significantly reduce the sublimation of the glaciers.

Wildfire Reduction

It is sad to see the wildfires all over the world, and wild animals burnt alive – the recent Australian wildfire consumed over a billion wild animals (Figure 7). It is time for climate scientists realize the importance of CO_2 and related GHG in the atmosphere and act upon it, which will substantially increase the humidity-holding capacity in the atmosphere. Further, CO_2 and GHG are natural fire retardants. It will lessen the chances of wildfires considerably. Also, the slower wind speed will help a lot.

Figure 7 Wild lives

Kuldip C. Tangri

Earth Is Getting Hotter

The earth is getting hotter. The atmospheric air is functioning like an automobile without a transmission. Even though the engine is running at full speed, the automobile is not moving. As previously mentioned, heat transfers by conduction, convection, and radiation. The conduction and radiation parts of heat transfer from the surface of the Earth are working, although they are not working 100%. The convection part of the heat removal system from the surface of the Earth is not working, where it accounts for 75% of heat removal. The scientific community knows that atmospheric air does not absorb heat. If it does not absorb heat, it cannot pluck the heat from the surface of the Earth and carry it to a higher altitude.

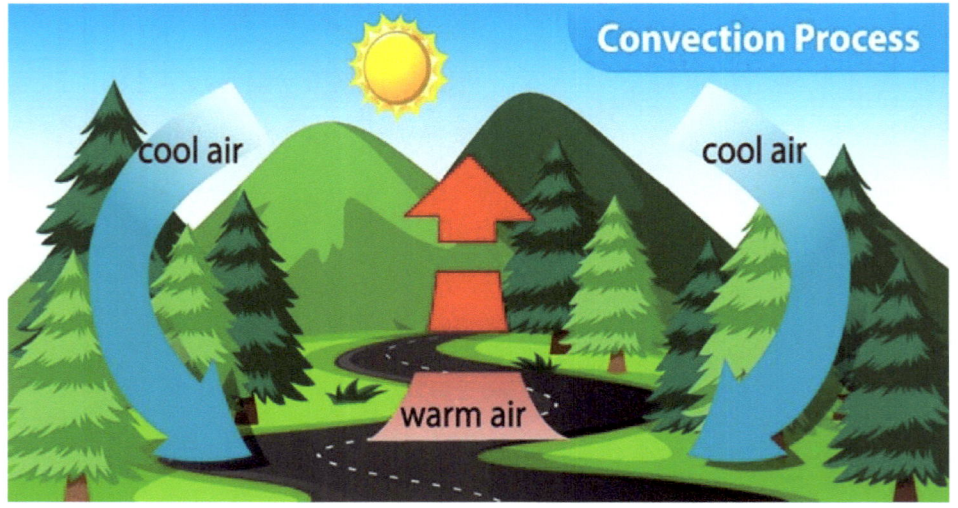

Figure 8

See the above Figure 8. The convection cooling system of Earth works exactly the way it is shown in the picture. The air cycle looks perfect. But this is not cooling the Earth; all it is doing is circulating the air. Air gets hot at the surface, goes up, cools off, and comes down. But it is not absorbing (capturing) any heat from the surface of the Earth to take with it. Therefore, there is a need for CO_2 and GHG to fortify the air so it can absorb the heat from the surface of the Earth and carry it to a high altitude to cool it. The fortified air will capture the heat from the surface of the Earth, take it to a higher altitude, get cold, and return to the surface of the Earth. This cooling cycle will keep on repeating. Once the proper mixture of atmospheric air, CO_2, and other GHG complex is achieved, the Earth will cool drastically. Conduction and radiation's part in the cooling of the Earth will also improve. CO_2 and GHG are multi-functional gases.

Sources of the Earth's Energy

The sun is the only outside source of energy for planet Earth. The Earth also produces energy due to the chemical reactions inside of it. For a chemical reaction to take place, the Earth's temperature must be above absolute zero. Absolute zero temperature is equivalent to –273.15°C (or –459.67°F). Where all molecular activities stop, this kind of temperature may be possible in laboratories. In the atmosphere, it could only happen either at the beginning or at the end of the universe. At that time, all activity stops. As we know, the atmospheric temperature is much higher than absolute zero, so the Earth is continually producing heat due to the chemical actions inside of it and getting energy (heat) from the sun. Therefore, the Earth needs constant cooling and protection from the sun.

The heat of the Earth produces approximately 90% of the water vapor in the atmosphere.

TED Talk

While I was searching for the answers to the above equation, I happened to watch a TED Talk on YouTube.

I watched Mr. James E. Hansen gave a TED Talk about global warming. Mr. Hansen was comparing the temperature of Earth to Venus. He was pointing out that the temperature of Venus is over 900°F. And it is kept hot by a thick carbon dioxide atmosphere, which was a result of the abundance of CO_2 in Venus's atmosphere. Mr. Hansen should know better. **Planet Venus is 25.72 million miles closer to the sun than the Earth**, and CO_2 has nothing to do with keeping the planet Venus hot. This man has a PhD. He was a director of NASA and he is a well-respected scientist in the climate change community. I believe it may be unintentional. Whatever Mr. Hansen and his team did in the 1980s, we are bearing the fruit now. Including his testimony before the U.S. Senate in 1988. I watched this video on YouTube. The hall was packed with people, and they were listening to Mr. Hansen, like a messiah was talking. (Link: https://youtu.be/fWInyaMWBY8).

Somebody is Cooking Books

In the 1960s automobile and industrial pollution in the atmosphere of American and European industrial cities was so high, people did not see the sun for months. At that time, there was a concern of the ice age. The sea levels were low, and polar bears were thriving. There were fewer downpours, floods, and wildfires, and the numbers of tropical depressions were way down. People were generally healthy. The big Kirloskar engines were running on oil in the factories, and industrial boilers were burning coal or oil. Worldwide railroads were using steam engines, which also used coal. There were plenty of very tall chimneys to exhaust the smoke. I do not understand how there could have been less CO_2 and GHG in the atmosphere during that period compared to now. It looks like somebody is cooking the books.

Volcanoes and Wildfires

They fulfill the need of nature for CO_2 and GHG. Before and during the Industrial Revolution, people burned coal, firewood, leaves, and even animal dung to meet their day-to-day need of cooking and warming their residences. Indirectly, they were helping nature to fulfill the need for CO_2 and GHG. Trees produce CO_2 at night. In case nature was still deficient in CO_2 and GHG, volcanoes and wildfires fulfilled the need for CO_2 and GHG. Volcanic eruptions were frequent. Industrial CO_2 and GHG started to fulfill the need for CO_2 and GHG in the Earth's atmosphere and the Earth started to get colder.

As a result of the cooling of the Earth, thousands of volcanoes went dormant. It depends on how much abuse nature can take. Eventually, it is going to reactivate the dormant volcanoes. Nature does not care about the Paris Agreement or the Kyoto Protocol. It is not going to be that simple to reactivation of dormant volcanoes. The size of the new volcano eruption is nothing compared to what they were. They can be so big that they can shoot out liquid nitrogen or any other gas hundreds of kilometers in the sky, and all organisms are going to burn, die, and freeze instantly. It is going to take years to thaw the Earth. There are many volcanoes in the works. At this time, volcanoes produce about 950 million tons of CO_2 and GHG a year. And nature is still short of CO_2.

There is a list on the volcano websites of volcanoes that erupted in 2019, and those can erupt again. There is a well-established volcano of Mount Etna, which erupts often. You should watch this video on YouTube. It shows the power of a volcanic eruption and shows how CO_2 is part of our daily life and all around us. The narrator did an outstanding job of narration. However, the narrator has a different opinion of CO_2 and the Earth's temperature, which I don't agree with because of the laws of thermodynamics, says different. Here is the link: https://youtu.be/6VUPIX7yEOM

The Climate Is Constantly Changing

The climate is constantly changing, but not drastically like it is now. Nature is a nurturer but has no problem making us uncomfortable or even killing us if we interfere with her routine. Nature can take extreme measures, such as inducing cyclones, droughts, earthquakes, floods, hail, hurricanes, lightning, rain, eruptions, snowstorms, thunderstorms, tornadoes, tsunamis, unbearable cold and heat, volcanoes, wildfires, windstorms, etc. Nature has all the above powers, so it does not have to make CO_2 and GHG to kill us or make us uncomfortable. The universe created CO_2 and all the other GHG to accommodate all the organisms on Earth, so we can survive and live healthy lives. As I mentioned previously, nature needs a certain amount of CO_2 and GHG to function every day. Any deficiency of these gases can kick in any of the above disasters. It looks like some of them are already in the works. And recently, one Australian wildfire.

CO_2 and GHG Are Crucial

They are as crucial to the Earth as feathers to a bird, leaves to a tree, water to a fish, oxygen for us, and amniotic fluid to a baby in the womb. It is the Earth that has survived the abuse of the pseudo-scientists.

Thanks to China and India

We owe our thanks to China and India for keeping up with the supply of CO_2 and GHG; otherwise, this destructive moment would have come to a lot sooner. If we do not add CO_2 and GHG to the atmosphere soon, things are possibly going to get worse. However, recent wildfires have pumped a few million megatons of CO_2 and GHG into the atmosphere of the Earth. Sadly, they will not last long, because these gases dissolve in water and will be used up quickly. The atmosphere is starving for these gases. Nature may be contemplating her next move to meet the ongoing need for these gases. Recent wildfires have produced some of CO_2 and related GHG. This may stabilize the weather for a while, but may not for long, so we still need to find a permanent solution.

Kyoto Protocol

The Kyoto Protocol has added a few man-made gases into the GHG category. They are hydrofluorocarbons (HFCs), perfluorocarbons (PFCs), and sulfur hexafluoride (SF_6), which are industrial gases.

Greenhouse Gases (GHG)

GHG or other GHG consist of many gases in the atmosphere, and they are less than 2% of all the atmospheric air. Most of these gases are in the troposphere. This chapter consists of the natural GHG which are in the atmosphere due to fossil fuel burning. Methane (CH_4) is an unburned fossil fuel and it is addressed accordingly. However, my focus is on the GHG which are in the atmosphere due to the combustion or burning of fossil fuels, predominantly CO_2. Their permitted quantities in the atmosphere need to be reassessed. These gases include, but are not limited to, ammonia (NH_3), carbon dioxide (CO_2), nitric oxide (NO), nitrogen dioxide (NO_2), nitrous oxide (N_2O), ozone (O_3), and sulfur dioxide (SO_2). A summary of the few notorious GHG starts with carbon dioxide.

Carbon Dioxide (CO_2)

Freezing point: -78.5°C (or -109.2°F).

Out of all the above gases, the most badmouthed gas is CO_2. All you hear about is a carbon footprint, reducing the carbon footprint, or creating a quota system for carbon. As a matter of fact, nowadays this is a mantra for financial gain and social ranking. All you have to do is talk unhappily about CO_2 and start talking about zero emissions. CO_2 does not store (hold) or produce any cold or heat; it just process the cold or heat. Similar to a microprocessor in a computer, it does not add or subtract any calculations; it just processes them. CO_2 decreases the effects of infrared radiation and protects the Earth from the radiation of the sun. It is less than 0.035% of the entire atmospheric air. It can be as high as 0.065% of the atmospheric air without having any significant consequences. We use it in soft drinks by the tons. There is no such thing a footprint.

CO_2 has nothing to do with global warming, but the deficiency of CO_2 does. Civilization and wildfires have wiped out more than 75% of the tree population in the world. The loss of these trees has significantly affected the supply of CO_2 and oxygen. CO_2 is an opaque color and a noncombustible gas; it is a natural fire retardant. It also increases the moisture-holding capacity of the atmospheric air at a lower temperature while reducing rapid combustion. Lately, most deforestations are done by wildfires, due to the lack of humidity in the atmosphere. CO_2 absorbs and disperses heat. That is what makes it an essential material on the Earth for the climate system. It has an acidic taste and has a distinct odor. The odor and taste of this gas are undetectable because the quantity of this gas in the atmosphere is meager.

There are many uses of CO_2, starting with food and the medical industry, which use CO_2 to preserve their ingredients. Dry ice is a form of CO_2, and it is also a food preservative. The beer and soft drinks industry use CO_2 in large quantities. However, the largest consumer of CO_2 is nature. On any given day, nature needs at least 30 million megatons of CO_2 to function halfway decently. The life cycle of CO_2 is concise. CO_2 is with us all the time, from the day we born until the day we depart. It runs through our veins. This is a motherly gas. The climate system cannot function without this gas. CO_2 is essential for rock formations, clamshells, the bones, and hard-shell organisms, which all need CO_2. Right now, sea creatures are dying or being deformed due to the deficiency of CO_2.

Methane (CH_4)

Freezing point: -182°C (or -295.6°F).

Methane is a colorless, odorless, and flammable gas. It should not fall in the category of GHG because it is a live fossil fuel. Nature produces it in large quantities, without the help of either animals or humans. Nature uses lightning to get rid of escaped methane and other unwanted flammable gases in the atmosphere. It is not soluble in water, but the residue after the burning of this fuel is water soluble. Why is methane considered a GHG? It is a live fuel. Most GHG are the byproducts of fossil fuels burning and are water soluble, whereas methane is not.

Nitrous Oxide (N_2O)

Freezing point: -90.86°C (or -131.5°F).

Nitrous oxide is a colorless and non-flammable gas. It is called laughing gas. Most organisms, all the waste products, fields, streams, and oceans, produce it. I spent hundreds of hours watching these climate change professors and scientists on YouTube and the Internet all talking badly about N_2O. This gas is water-soluble. Therefore, its life in the atmosphere is limited. It has been ridiculed just like any other GHG by scientists and nonscientists and accused of being 300 times worse than CO_2 in the matter of global warming. N_2O is next to CO_2 on the assassination list of the Kyoto Protocol. The deficiency of this gas is detrimental to farming, vegetation, and organisms. This gas is a component of the clouds, and the height of the clouds depends on the combination of the GHG and water vapor. These clouds turn into rain, snow, and any other available methods to distribute the trace of this gas to vegetation and organisms. It is therefore a source of amino acids, proteins, and required nutrition that most organisms need. N_2O can now be produced on demand. This gas is an integral part of the climate system.

Nitrogen Dioxide (NO_2)

Freezing point: -11°C (or 11.84°F).

I know nitrogen dioxide is on the list for causing acid rain. It is corrosive, and the safe quantity in the atmosphere should be rechecked and increased to the allowable amount if possible. NO_2 not only sustains the clouds, but also serves as a conduit to deliver amino acids, protein, DNA, RNA, and other growth factors to organisms. All living organisms need a trace of this gas.

Sulfur Dioxide (SO$_2$)

Freezing point: -72°C (or -97.6°F).

Sulfur dioxide is a poisonous gas, and it is recommended that it should not be inhaled. Besides that, it helps with the clouds, and all organisms need the element of sulfur. Nature delivers this element to organisms through different conduits, and one of them is sulfur dioxide. Without this element, organisms will die a cruel death. We do not have to go to the sulfur mines to get this element, as nature delivers it by air and water through vegetation and food. The Industrial Revolution produced GHG that helped with the delivery of this mineral – and other minerals – to organisms indirectly, through fruits, grains, and vegetation. Of course, excess of everything is bad, so we need to manage these gases, not eradicate them.

Atmospheric Layers

The atmosphere of the Earth is different compared to the other planets. It has five distinctive layers, as shown here in Figure 9, and they are called the troposphere, stratosphere, mesosphere, thermosphere, and exosphere. The temperatures inside each layer are different and explained here.

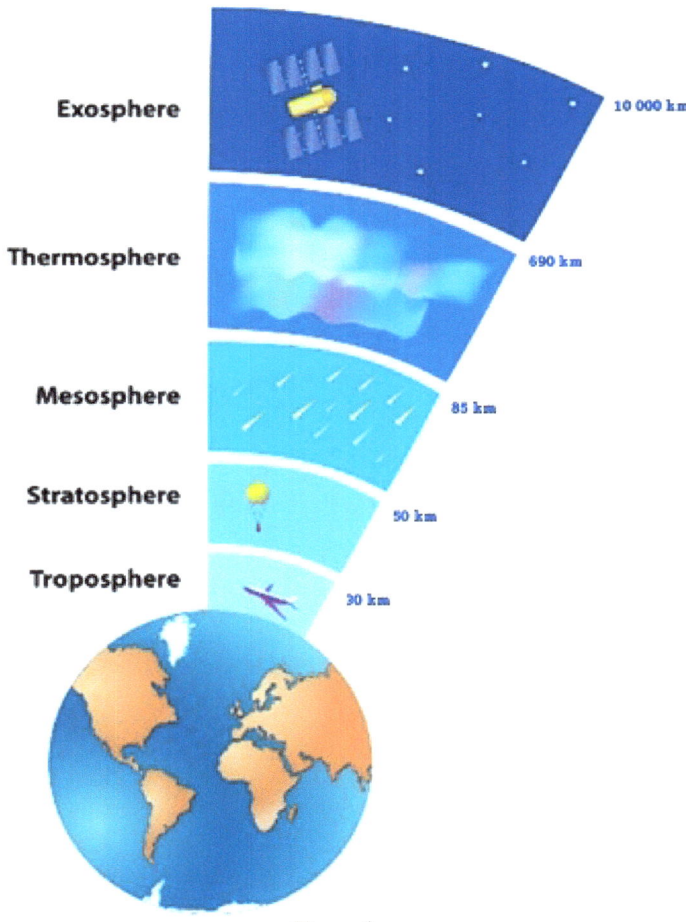

Figure 9

The Troposphere

Altitude: 0 to 15 km (0 to 9.32 miles).

The troposphere is the first atmospheric layer of the Earth. It begins at the surface of the Earth and extends vertically. This extension varies and is higher at the Equator and lesser at the poles. The temperature drops with height in this layer, which is called the lapse rate. The temperature drops with the elevation from the surface of the Earth. Sometimes this does not happen because there are many moving parts in the atmosphere, such as not enough CO_2 or GHG, etc. It is approximately 6.5°C per kilometer (or 18.8°F per mile). The troposphere is warmer at the bottom (the surface of the Earth) and cooler at the top (the bottom of the tropopause). Heat travels from the surface of the Earth toward the tropopause. Temperature inversion takes place in this layer. The tropopause is a buffer zone between the troposphere and stratosphere. The tropopause separates the troposphere and the stratosphere.

The troposphere is the layer where all the human, organism, and weather-related activities take place. CO_2, GHG, clouds, floods, hurricanes, volcanoes, water vapor, wildfires, and windstorms are a part of this layer. There is ongoing misinformation floating around that heat is trapped in this layer due to GHG and troposphere is a greenhouse, but it is not.

The Stratosphere

Altitude: 15 to 50 km (31.07 miles).

The stratosphere is the second-lowest atmospheric layer of the Earth. It extends vertically above the tropopause. This extension varies, and it is higher at the Equator and lesser at the poles. Above the stratosphere is the stratopause. The stratopause separates the stratosphere and the mesosphere. The ozone layer is right below the stratopause and above the stratosphere. This ozone layer works as a clear shield or canopy. This shield gets hot, due to its closeness to the sun, and it blocks the ultraviolet rays of the sun. Compared to the troposphere, the stratosphere layer is the opposite of the troposphere. It is colder at the bottom and hotter at the top. Airplanes fly in this layer.

It is a stable layer of the atmosphere. Sometimes, volcano ashes and high clouds penetrate the tropopause and shoot into this layer.

The Mesosphere

Altitude: 50 to 80 km (31.7 to 50 miles).

The mesosphere is the third-lowest atmospheric layer of the Earth. It extends vertically from the surface of the stratopause. This extension varies like other layers, and is higher at the Equator and lesser at the poles. In this layer, the temperature drops with elevation, up to the lower boundary of the mesopause. The mesopause separates the mesosphere and the thermosphere. Below the mesopause, the air is frigid, and the water vapor at this elevation gets sublimated into polar-mesospheric clouds or noctilucent clouds. These are the highest clouds in the atmosphere. It is an ionosphere region of the atmosphere.

The Thermosphere

Altitude: 80 to 700 km (50 to 440 miles).

The thermosphere is the fourth-lowest layer of Earth's atmosphere. It extends from the top of the mesopause to the bottom of the thermopause. The distance between mesopause and thermopause fluctuates due to the changes in solar activity. This layer is usually extremely hot, with temperatures reaching above 1600°C. The air in this layer is very thin. The International Space Station orbits in this layer.

The Exosphere

Altitude: 700 to 10,000 km (440 to 6,200 miles).

The exospheric layer is above the thermosphere layer of the atmosphere of the Earth. It extends from the top of the thermopause. This layer has traces of all the gases, but helium and hydrogen are the leading gases. Satellites and spacecraft are residents of this layer. The days are very hot, and nights are very cold in this layer.

Clouds

There are three ingredients needed to make a cloud: water vapor, greenhouse gases, and atmospheric air to carry and move them around. There are dust clouds in the atmosphere, but they are not common. Getting water vapor from the ocean and carrying it to the mountain peaks is not straightforward. There is a science involved in this process. Pure water vapor clouds will collapse prematurely and turn into a downpour. Nature mixes CO_2 and related GHG with water vapor in the atmosphere to lower the freezing point of water vapor. Clouds made of this mixture travel farther and higher in altitude without falling apart in the atmosphere.

Types of Clouds

There are many types of clouds and a few are listed here. The height of these clouds in the atmosphere depends on the quantity and the type of GHG and water vapor.

Low-Level Clouds

Cumulus clouds, max elevation 1.5 km (4,921.26 ft).

Stratus Clouds

Max elevation 2 km (6,561.68 ft).

Stratocumulus Clouds

Max elevation 2.5 km (8,202.1 ft).

Arcus Clouds

Low and horizontal clouds associated with cumulonimbus.

Mid-Level Clouds

Nimbostratus clouds, max elevation 3.51 km (11,500 ft).

High-Level Clouds

Altocumulus clouds, max elevation 6.5 km (21,325.46 ft).

Altostratus clouds, max elevation 6.5 km (21,325.46 ft).

Cirrus clouds, elevation above 5.7 km (18,700.79 ft).

Cirrocumulus clouds, elevation 6.1-12.5 km (20,000-41,000 ft).

Cirrostratus clouds, elevation 6.1-13 km (20,000-42,650 ft).

Cumulonimbus Clouds

2-18.5 km (2,500-60,700 ft) and classified as a vertical cloud.

Pyrocumulus Cloud

Also known as a fire cloud. A dense cumuliform cloud associated with fire or volcanic eruptions.

The Intergovernmental Panel

According to the Intergovernmental Panel On Climate Change of the UN, carbon dioxide (CO_2) absorbs and radiates heat. So what? Everything absorbs and disperses heat. Animals absorb heat, buildings absorb heat, canopies absorb heat, and the Earth absorbs heat. Furthermore, helmets absorb heat, humans absorb heat, tents absorb heat, and umbrellas absorb heat, and they all disperse heat. In addition, they provide their respective users with the expected comfort. Anything between Earth and the sun protects the Earth from sun rays, including CO_2 and GHG. The quality of protection depends on the percentage of CO_2 or and GHG in the atmosphere air. We use sunglasses to reduce the effect of sunlight on our eyes; likewise, greenhouse gases reduce the impact of the sun's rays on the Earth.

The Pseudoscience

Looks and sounds like science, but it is not science. The scientists of the World Meteorological Organization and related organizations on climate system have worked for the last 68+ years to eradicate CO_2 and other GHG. Their intentions were sincere. But the information they were acting upon was not scientifically proven. And they are still promoting the same unproven information. Since the early 1980s; the world climate has started to show the signs of gradual changes in the form of desertification, drought, floods, glaciers melting, destructive hurricanes, mountain peaks losing snow, sea levels rising, wildfires, and damaging wind speeds. Instead of reviewing their science, the scientists kept on blaming CO_2 and GHG. Sadly, the scientists of the EPA, WMO, IPCC, UNFCCC, and all the universities of the world are still blaming CO_2 for global warming and climate change. I believe because CO_2 absorbs, disperses, and radiates heat. CO_2 indeed absorbs and disperses heat, but what does the heat absorption of CO_2 have to do with global warming and climate change? It means it is a material that is a good conductor of heat, like metals.

Heatsinks

Figure 10

Look at the above Figure 10 of Heatsinks in use. They capture the heat from the surface of electronic parts in computers or machines and disperse the heat into the atmosphere. They are made of a metal that is a good conductor of heat. Similarly, that is what CO_2 and related GHG do to cool the Earth. They mix with the atmosphere, air, which enables the atmospheric air to capture heat from the surface of the Earth. It does not need a large quantity of CO_2 to fortify the atmospheric air. Once a proper mixture of CO_2 and atmosphere, air is achieved, atmospheric air becomes a heatsinker, where it captures heat from the surface of the Earth and cools it by convection. In case of a computer or machine depending on the size, sometimes fans are used to further disperse the heat from the Heatsinks.

Suitability of Material

All materials have different properties, such as aluminum, buildings, copper, CO_2, water vapor, parchment paper, toilet paper, etc. You use a specific material for a certain job due to the suitability of the material. The suitability depends on the properties of the material. Engineers and scientists choose materials for a project according to the properties of a material. For example, in a project, an engineer must use a metal that is a good conductor of electricity and heat. It must tolerate higher temperatures. Where money is no issue, the engineer checks the properties of all different metals. Out of all the metals, the most suitable metal is copper. Therefore, he uses copper. Similarly, if a person wants a balloon to go up, he fills it with helium.

Parchment Paper

In choosing between parchment paper or toilet paper, we use toilet paper. Why not use parchment paper instead of toilet paper? Because parchment paper does not have the properties of toilet paper, and it is not suitable for the job of toilet paper. The same goes for the atmospheric air without CO_2 or GHG, which is like parchment paper because it will not absorb (capture) heat from the surface of the Earth, and transport it to higher altitudes to cool the Earth by convection; 75% of the Earth is cooled by convection.

Adam the Project Engineer

Adam was the project engineer. He was required to transport the heat from the Equator to the North Pole. He studied the project and realized that he needed a method of transportation that carries heat. He looked at his options and decided that atmospheric air should be fine and should move the heat from the Equator to the North Pole in no time. The project started, and it has been going on for a few days. Air is moving; everything looks okay, but the heat is not moving. He searches for the answer: Why is it not working? He knew the properties of the materials and realized that atmospheric air does not absorb heat; therefore, it cannot capture and carry heat. Being an engineer, he knew how to fix this problem. He fortified the atmosphere, air with CO_2 and related GHG to fix the problem. He knew exactly how much CO_2 and suitable GHG to add to fortify the atmospheric air so it could absorb heat. Once he attained the proper blend of CO_2 and GHG to fortify the atmospheric air, his transportation of heat started to work. He did not need to add a lot of CO_2 to the atmospheric air to fortify it: 0.0450% of CO_2 was enough. Nature does it automatically all the time. Nature is the environmental and chemical engineer of the universe.

The average CO_2 in the atmospheric air is less than 0.0370%, not to account for the glut of CO_2 in China and India. It can go as high as 0.0650% with no drastic consequences. The only effect of higher levels of

CO_2 would be that the temperature of the Earth and sea level may go down faster. Also wind speed would be reduced drastically.

Properties of Air

Gas, liquid, and solid: all materials, including air, greenhouse gases, liquids, and solids, are thermally conductive and have density according to their content identity. They all absorb, conduct, and emit heat until the temperature drops to absolute zero (-273°C or -459°F), where all the activity stops.

Water Vapor

The Earth is producing energy and getting energy from the sun. Therefore, the Earth needs constant cooling and protection from the sun. The Earth's heat produces about 90% of the water vapor in the atmosphere. Water evaporates either from ponds, rivers, or the seas. It cools the Earth and evaporated water piggybacks the atmosphere, air like any other gas or like dust particles. The life cycle of water vapor in the atmosphere depends on air temperature, the quantity of greenhouse gases in the atmosphere to fortify it, and the height in altitude.

The Climate

Many components make up the climate. Like automobiles, buildings, human bodies, etc. Any single component can affect the performance of the subject. The same goes for the climate. The importance of CO_2 and related GHG in the atmosphere has been established in previous chapters. Therefore, it is important that we must reintroduce these excessively removed components back into the atmosphere, so the climate can function properly. The sooner the proper blend of air and CO_2 and related GHG is achieved, the sooner things will fall in place. The Earth will start to cool, and sea levels will start to go down drastically. There is going to be a domino effect of the benefits.

Janicki Omni Processor

Mr. Bill Gates created the Janicki Omni processor to distill poop water. The poop water processor looked very impressive. It almost looked like a very expensive moonshine machine. Mr. Gates wanted to solve the drinking water problem. I believe he has realized it is not that easy.

Here is a link to the Janicki Omni processor:

https://youtu.be/p-UaKGISXjQ.

Nature Is a Mighty Distiller

Nature is the mighty distiller of the world. Who distills the water from a pond, river, or sea 24/7, then separates all impurities from the water and transports it all over the world reservoirs to store it? It further adds minerals and nutrients while transporting the distilled water to the proper destinations. CO_2 and GHG are integral parts of the distillation process of nature. They stay in this process from the time the operation begins to when it ends. The operation ends when distilled water reaches the destination. The goods are delivered, and the containers (CO_2 and GHG) surrender with the goods. They leave the atmospheric air fresh and clean.

Abbreviations

Ar Argon

CO_2 Carbon dioxide

GHG Greenhouse gases

He Helium

H_2O Water

H Hydrogen

CH_4 Methane

N_2 Nitrogen

NH_3 Ammonia

NO Nitric oxide

NO_2 Nitrogen dioxide

N_2O Nitrous oxide

N_2 Nitrogen

O_3 Ozone

O_2 Oxygen

SO_2 Sulfur dioxide

EPA – Environmental Protection Agency

GPM - Global Precipitation Measurement

IPCC – Intergovernmental Panel on Climate Change

JAEA - Japan Aerospace Exploration Agency

TD – Tropical depressions

NASA – The National Aeronautics and Space Administration

UN – United Nations

UNFCCC – United Nations Framework Convention on Climate Change

UNO – United Nations Organization

WMO – World Meteorological Organization

CO_2 is an integral part of the climate system.

CO_2 is a motherly gas that runs through our veins.

CO_2 does not cause global warming.

CO_2 is a refrigerant of the atmosphere.

The lifespan of CO_2 is limited in the atmosphere.

How can we reduce the Earth's temperature?

How can we reduce the number of earthquakes?

How can we reduce desertification?

How can we reduce downpours?

How can we increase the humidity in the atmosphere without increasing the temperature?

How can we lower the sea level?

How can we reduce the volcanic eruptions?

How can we reduce the wildfires?

How can we reduce the wind speed?

www.ingramcontent.com/pod-product-compliance
Lightning Source LLC
Chambersburg PA
CBHW040226220526
45473CB00001B/139